AF095778

# THE POETRY OF HOLMIUM

# The Poetry of Holmium

Walter the Educator

Silent King Books a WhichHead Imprint

Copyright © 2024 by Walter the Educator

All rights reserved. No part of this book may be reproduced in any manner whatsoever without written permission except in the case of brief quotations embodied in critical articles and reviews.

First Printing, 2024

Disclaimer
This book is a literary work; poems are not about specific persons, locations, situations, and/or circumstances unless mentioned in a historical context. This book is for entertainment and informational purposes only. The author and publisher offer this information without warranties expressed or implied. No matter the grounds, neither the author nor the publisher will be accountable for any losses, injuries, or other damages caused by the reader's use of this book. The use of this book acknowledges an understanding and acceptance of this disclaimer.

"Earning a degree in chemistry changed my life!"
- Walter the Educator

dedicated to all the chemistry lovers, like myself, across the world

# CONTENTS

Dedication . . . . . . . . . . . v

Why I Created This Book? . . . . . . . . . 1

**One** - Oh Holmium . . . . . . . . . 2

**Two** - The Story Of Holmium . . . . . . . . 4

**Three** - Enigmatic And Bold . . . . . . . . 6

**Four** - Cosmic Conductor . . . . . . . . 8

**Five** - Brilliance And Awe . . . . . . . . . 10

**Six** - Forever Cherished . . . . . . . . . . 12

**Seven** - Symphony So Grand . . . . . . . 14

**Eight** - Knowledge Profound . . . . . . . . 16

**Nine** - Hidden From Sight . . . . . . . 18

**Ten** - Gem Of The Periodic Table . . . . . 20

**Eleven** - Saga Of Holmium . . . . . . . . . 22

**Twelve** - Grand Narrative . . . . . . . . . 24

**Thirteen** - Element Of Lore . . . . . . . . . . 26

**Fourteen** - Quantum Chase . . . . . . . . . . 28

**Fifteen** - Symbol Of Progress . . . . . . . . 30

**Sixteen** - Never Dims . . . . . . . . . . . . 32

**Seventeen** - Presence Is Profound . . . . . . 34

**Eighteen** - Treasure To Adore . . . . . . . . 36

**Nineteen** - Inspire With Your Complexity . . 38

**Twenty** - Scientists Investigate . . . . . . . . 40

**Twenty-One** - Guardian Of Dreams . . . . . 42

**Twenty-Two** - Future So Bright . . . . . . . . 44

**Twenty-Three** - Fiber Optics To Lasers . . . . 46

**Twenty-Four** - Healing The Sound . . . . . . . 48

**Twenty-Five** - Mystery Profound . . . . . . . 50

**Twenty-Six** - Divine . . . . . . . . . . . . . 52

**Twenty-Seven** - Science Redeems . . . . . . . 54

**Twenty-Eight** - Discoveries Galore . . . . . . 56

**Twenty-Nine** - Spectral Dance . . . . . . . . 58

**Thirty** - Atomic Art . . . . . . . . . . . . . 60

**Thirty-One** - Corners Of The Earth . . . . . 62

**Thirty-Two** - Metal With Allure . . . . . . . . 64

**Thirty-Three** - Luminescent Hue . . . . . . . . 66

**Thirty-Four** - Symbol Of Innovation . . . . . 68

About The Author . . . . . . . . . . . . . . . 70

# WHY I CREATED THIS BOOK?

Creating a poetry book about the chemical element Holmium was an innovative way to explore its properties, generate interest in lesser-known elements, and foster interdisciplinary connections between science and art. It allows for a creative expression of the element's essence and can captivate readers with its unique blend of scientific knowledge and poetic beauty.

# ONE

# OH HOLMIUM

In the realm of elements, a tale unfolds,
Of Holmium, a metal with tales untold.
With atomic number sixty-seven,
Its mysteries reside in heavens' given.

A rare earth element, in nature's embrace,
Holmium shines with a captivating grace.
Its lustrous beauty, a radiant gleam,
A cosmic dance, like a celestial dream.

Within its core, a magnetic force,
Holmium's power, a magnetic discourse.
It aligns its spins, in perfect array,
A magnetism that holds darkness at bay.

From the laboratories to the depths below,
Holmium's uses, like rivers that flow.

In lasers and optical fibers it resides,
Guiding us through the unseen divides.

Through the ages, its story unfolds,
As Holmium's secrets, the universe holds.
A symbol of strength, of science and art,
Unveiling the wonders that lie in our heart.

So let us marvel at Holmium's might,
A catalyst for innovation and light.
In the cosmos of elements, it takes its stand,
A testament to the wonders of the grand.

Oh Holmium, element of mystery and charm,
In your presence, we find solace and calm.
May your essence guide us, forevermore,
As we explore the vast unknowns that lie in store.

# TWO

# THE STORY OF HOLMIUM

In the realm of elements, a tale to be told,
Of Holmium, a metal with secrets unfold.
With sixty-seven protons, it takes its place,
A cosmic enigma, full of grace.

Holmium, the guardian of magnetic might,
With its magnetic force, it sets things right.
Aligning the spins, a symphony of power,
Invisible threads, weaving hour by hour.

A rare earth element, rare and divine,
Holmium's presence, a celestial sign.
Its lustrous aura, a shimmering glow,
A dance of electrons, in patterns that flow.

From lasers to magnets, its purpose is vast,
Holmium's allure, destined to last.

In optic fibers, it finds its embrace,
Guiding us forward, through time and space.

Oh Holmium, element of wonder and awe,
In your essence, mysteries we draw.
A symbol of strength, of knowledge untold,
Revealing the secrets the universe holds.

So let us celebrate Holmium's gift,
A catalyst for discovery and uplift.
In the realm of elements, it shines bright,
A beacon of science, a guiding light.

Holmium, the guardian of secrets unknown,
In your presence, inspiration is sown.
With every discovery, a new chapter unfolds,
In the story of Holmium, forever it molds.

# THREE

# ENIGMATIC AND BOLD

In the realm of elements, a tale unfurls,
Of Holmium, a metal that mesmerizes and swirls.
With sixty-seven protons, it claims its place,
A cosmic enigma, full of grace.

Holmium, the luminary of the rare earth clan,
With magnetic prowess, a celestial plan.
Its spins align, in a symphony of might,
Creating a force that dazzles in the night.

A conductor of lasers, it takes the lead,
Guiding us through realms, where knowledge we feed.
In fibers of optics, it weaves a path,
Illuminating our world, dispelling the dark.

Oh Holmium, the guardian of our dreams,
In your presence, hope forever gleams.

A symbol of resilience, strength, and might,
Unveiling the mysteries hidden from sight.

From laboratories to far-reaching skies,
Holmium's wonders, a sight for our eyes.
In alloys and magnets, its purpose is found,
Revolutionizing technology, breaking new ground.

So let us marvel at Holmium's allure,
A catalyst for progress, forever secure.
In the realm of elements, it stands tall,
A testament to the wonders of it all.

Holmium, the enigmatic and bold,
In your essence, stories of science unfold.
With every discovery, a legacy is paved,
In the tapestry of elements, you're engraved.

# FOUR

## COSMIC CONDUCTOR

In the cosmic symphony, a metal divine,
Holmium, a marvel, in treasures it does shine.
With sixty-seven protons, it takes its stage,
A magnetic enigma, scripting secrets on each page.

Holmium, the guardian of spectral delight,
In its atomic dance, colors ignite.
From vibrant greens to deep reds it creates,
A palette of hues, an artist it emulates.

A rare earth element, a gem in disguise,
Holmium's allure, a celestial prize.
Its lustrous sheen, like stardust aglow,
Weaving tales of wonder in the cosmos's flow.

In lasers it dances, a guiding star,
Illuminating paths, near and far.

Fiber-optic marvels, it breathes life,
Connecting worlds, diminishing strife.

Oh Holmium, element of mystique and grace,
In your presence, knowledge finds its place.
A symbol of exploration, innovation's creed,
Unveiling the depths where discoveries breed.

From laboratories to technological feats,
Holmium's essence, a symphony that repeats.
In magnets and alloys, it finds its call,
Revolutionizing industries, standing tall.

So let us marvel at Holmium's embrace,
A catalyst for progress, leaving a trace.
In the grand tapestry of elements, it thrives,
A testament to the wonders of our lives.

Holmium, the cosmic conductor we adore,
In your essence, the universe we explore.
With every revelation, a new chapter takes flight,
In the eternal saga of Holmium's light.

# FIVE

# BRILLIANCE AND AWE

Holmium, element of ethereal might,
In the realm of science, a shining light.
With atomic beauty, you captivate,
A cosmic symphony, orchestrating fate.

Your magnetic force, a celestial dance,
Aligning the spins in a cosmic trance.
In lasers and fibers, your purpose unfolds,
Guiding us through mysteries yet untold.

Oh Holmium, guardian of secrets profound,
In your essence, hidden wonders are found.
A symbol of strength, resilience, and grace,
Unveiling the universe's intricate embrace.

From laboratories to technological strides,
Holmium's power, the world it guides.

In magnets and alloys, it finds its place,
Revolutionizing industries with its embrace.

So let us marvel at Holmium's gleam,
A catalyst for progress, a visionary's dream.
In the tapestry of elements, you shine bright,
Igniting curiosity, igniting our sight.

Holmium, element of brilliance and awe,
In your presence, boundaries withdraw.
With every breakthrough, a new frontier,
In the saga of Holmium, forever we steer.

# SIX

## FOREVER CHERISHED

Oh Holmium, a treasure embraced by the Earth,
With your magnetic charm, you captivate since birth.
Symbol of strength, a cosmic embodiment,
Unveiling mysteries with every experiment.

In laboratories, your essence takes flight,
Guiding us through the darkness, igniting the light.
From lasers to optical fibers, your purpose so clear,
Leading us forward, dispelling all fear.

A guardian of knowledge, you stand tall,
Revealing the secrets hidden within your thrall.
In magnets and alloys, your power prevails,
Revolutionizing industries, pushing the scales.

Oh Holmium, element of wonder and might,
In your presence, discoveries take flight.

A catalyst for progress, a scientific muse,
Inspiring innovation, breaking through the clues.
   So let us celebrate your elemental grace,
A testament to the wonders of time and space.
In the grand tapestry of elements, you're adored,
Holmium, forever cherished, forever explored.

# SEVEN

# SYMPHONY SO GRAND

In the realm of elements, Holmium stands proud,
With its secrets and mysteries, it speaks aloud.
A symbol of strength, resilience, and might,
Unveiling the wonders hidden from sight.
    Oh Holmium, element of rare allure,
In your essence, curiosity finds its cure.
From lasers to magnets, your purpose unfolds,
Revolutionizing technologies, pushing new thresholds.
    Through laboratories, your brilliance is sought,
Guiding us forward, with knowledge you've brought.
In alloys and fibers, your presence is known,
A catalyst for progress, in realms yet unshown.
    So let us marvel at Holmium's embrace,
A beacon of science, a celestial grace.

In the cosmic dance of elements, you shine,
Igniting the flames of discovery, intertwined.
  Holmium, the enigmatic and profound,
In your essence, exploration is found.
With every breakthrough, a symphony so grand,
In the story of Holmium, forever we stand.

# EIGHT

# KNOWLEDGE PROFOUND

In the realm of elements, Holmium shines,
A mystical presence, so rare and refined.
Symbol of strength, with secrets untold,
Unveiling mysteries that science behold.
    Oh, Holmium, element of enchantment and might,
In your essence, wonder takes flight.
With every discovery, a new door opens wide,
In the cosmic symphony, you beautifully reside.
    From laboratories to technological feats,
Holmium's power, the world it greets.
In magnets and lasers, your purpose is clear,
Revolutionizing progress, breaking barriers near.
    So let us celebrate Holmium's allure,
A catalyst for innovation, forever secure.

In the tapestry of elements, you weave,
A testament to the wonders we perceive.
   Holmium, the guardian of knowledge profound,
In your presence, wisdom is found.
With every revelation, a new chapter unfurls,
In the story of Holmium, where brilliance whirls.

# NINE

# HIDDEN FROM SIGHT

Holmium, element of radiance and grace,
In your atomic dance, mysteries embrace.
A symbol of strength, resilient and bright,
Unveiling the secrets hidden from sight.

Through laboratories, your wonders are sought,
Guiding us forward, where knowledge is wrought.
In magnets and lasers, your power unfolds,
Revolutionizing industries, forging new molds.

Oh Holmium, conductor of energy's flow,
In your presence, innovation will grow.
A catalyst for progress, a visionary's delight,
Igniting the spark that illuminates the night.

So let us marvel at Holmium's allure,
A beacon of science, forever pure.

In the grand tapestry of elements, you're revered,
A testament to exploration, never to be veered.
 Holmium, the cosmic storyteller untold,
In your essence, the universe does unfold.
With every discovery, a new chapter unfurls,
In the epic saga of Holmium, where knowledge swirls.

# TEN

# GEM OF THE PERIODIC TABLE

Holmium, a gem of the periodic table,
With properties unique, it's truly able.
In laboratories, its secrets unfurl,
A testament to science, an alchemical swirl.

Magnetic allure runs through its core,
Harnessing power like never before.
In magnets and lasers, its presence is felt,
Revolutionizing technologies, making them excel.

Oh Holmium, conductor of light,
In your brilliance, innovation takes flight.
A catalyst for progress, a visionary's dream,
Igniting the spark of discovery's gleam.

Let us marvel at Holmium's mystique,
A silent hero, humble and meek.

In the grand tapestry of elements, you stand,
A symbol of human intellect, hand in hand.
Holmium, the guardian of knowledge's quest,
In your essence, wisdom manifests.
With every revelation, a new frontier is crossed,
In the epic story of Holmium, forever embossed.

# ELEVEN

# SAGA OF HOLMIUM

In the realm of elements, Holmium sings,
A symphony of atoms, where knowledge springs.
With atomic number sixty-seven, it stands,
A conductor of science, in skilled hands.

Holmium, the guardian of magnetic might,
In your presence, mysteries take flight.
In magnets and lasers, your power blooms,
Revolutionizing technology, breaking old rooms.

Oh Holmium, the luminescent gem,
In your brilliance, wonders never condemn.
A catalyst for progress, an inventor's spark,
Igniting innovation, leaving an indelible mark.

Let us celebrate your atomic grace,
An element of brilliance, a cosmic embrace.

In the tapestry of elements, you shine,
Holmium, a testament to the divine.
From laboratories to vast research halls,
Holmium's secrets, the curious enthralls.
In alloys and fibers, its purpose unfurls,
Unleashing the potential of scientific worlds.
Holmium, the enigma of the periodic chart,
In your presence, discoveries impart.
With every breakthrough, a new chapter unfolds,
In the saga of Holmium, where knowledge beholds.

# TWELVE

# GRAND NARRATIVE

Holmium, oh element rare and true,
In the realm of science, we turn to you.
With your magnetic charm, you captivate,
Leading us through mysteries, we contemplate.

In lasers and magnets, your power resides,
Revolutionizing technologies, pushing tides.
A catalyst for progress, you pave the way,
Unveiling secrets, day by day.

Let us marvel at your atomic dance,
A symphony of particles, a cosmic chance.
In the tapestry of elements, you're distinct,
Holmium, a gem we must not neglect.

From laboratories to industry's core,
Your presence is felt, forevermore.

In alloys and fibers, you leave your trace,
Advancing our world at a rapid pace.
    Holmium, the guardian of knowledge's door,
In your essence, wisdom does pour.
With every discovery, a new horizon unfolds,
In the grand narrative of Holmium, untold.

# THIRTEEN

# ELEMENT OF LORE

Oh, Holmium, element of lore,
In your essence, brilliance does pour.
From laboratories to scientific quests,
You unveil the secrets, putting minds to the test.

In magnets and lasers, your power resides,
Revolutionizing technology's strides.
A catalyst for progress, innovation's guide,
Unlocking the mysteries, side by side.

Let us marvel at your atomic might,
A beacon of knowledge, shining so bright.
In the grand tapestry of elements, you stand,
Holmium, the catalyst of human demand.

From alloys to fibers, you leave your mark,
Advancing industries, lighting the spark.

In the symphony of science, you play your part,
Unveiling the wonders, igniting the heart.
   Holmium, the guardian of intellectual chase,
In your presence, discoveries embrace.
With every revelation, a new chapter unfurls,
In the epic saga of Holmium, where wisdom swirls.

# FOURTEEN

# QUANTUM CHASE

In the realm of colors, Holmium weaves its spell,
Infusing cubic zirconia, a beauty to compel.
With magnetic might, it shines like no other,
A gem's allure enhanced, its essence to uncover.

Beyond the surface, a quest for the unseen,
Scientists delve into a realm so keen.
In search of the elusive, a magnetic monopole,
Holmium's strength guides their ambitious goal.

But Holmium's wonders don't end there,
In the medical realm, it shows its care.
From lasers to diagnostics, it lends a hand,
Advancing healthcare, a noble demand.

Oh Holmium, element of intrigue and might,
In your essence, the universe takes flight.
From magnets to gems, you leave your trace,
Unraveling mysteries, with elegance and grace.

Let poets sing of your magnetic embrace,
And scientists marvel at your quantum chase.
Holmium, a symbol of knowledge's tower,
Unveiling the secrets of cosmic power.

# FIFTEEN

# SYMBOL OF PROGRESS

In the realm of elements, Holmium stands tall,
A marvel of nature, captivating all.
Its atomic dance, a symphony of grace,
Unveiling the secrets of the cosmic space.

Holmium, conductor of energy's flow,
In magnets and lasers, your powers grow.
A catalyst for progress, innovation's muse,
Guiding scientists on their quest to choose.

Let us marvel at your luminescent glow,
A beacon of science, forever aglow.
In the tapestry of elements, you're revered,
A testament to exploration, never to be veered.

Holmium, the guardian of wisdom's quest,
In your presence, knowledge manifests.

With every discovery, a new horizon unfolds,
In the epic saga of Holmium, where brilliance beholds.
    From laboratories to industry's core,
Holmium's secrets, the curious explore.
In alloys and fibers, its purpose unfolds,
Igniting innovation, as the story of science unfolds.
    Oh Holmium, element of endless allure,
In your essence, the universe does endure.
A symbol of progress, a beacon of light,
Holmium, forever shining bright.

# SIXTEEN

# NEVER DIMS

Holmium, the element of vibrant hue,
In the realm of science, we turn to you.
With magnetic might, you captivate our gaze,
Unveiling wonders in mysterious ways.

From the depths of laboratories, you emerge,
A catalyst for innovation, a scientific surge.
In lasers and magnets, your power is sought,
Pushing boundaries, achieving what was once thought.

Oh Holmium, conductor of electromagnetic grace,
In your presence, technology finds its space.
A symphony of atoms, dancing in harmony,
Revealing truths of the universe's tapestry.

In medical fields, you lend a healing hand,
Diagnostic tools guided by your command.

With precision and accuracy, lives you save,
A testament to the power that you engrave.
    Holmium, the guardian of knowledge's door,
In your essence, discoveries forever soar.
With every breakthrough, a new era begins,
In the grand saga of Holmium, where wisdom never dims.

# SEVENTEEN

# PRESENCE IS PROFOUND

Holmium, a luminary in the realm of science,
You captivate our minds with your magnetic alliance.
In the periodic table, you claim your rightful place,
A beacon of knowledge, radiating with grace.

With atomic prowess, you shape the world we know,
Unveiling mysteries, like a cosmic shadow.
In lasers and optics, you find your domain,
Guiding innovation, pushing the boundaries of the mundane.

Oh Holmium, conductor of electromagnetic might,
You illuminate the path, shining so bright.
In medical applications, your presence is profound,
Enhancing diagnostics, where hope is found.

From laboratories to industrial endeavors,

Your influence spreads, like ripples in rivers.
A catalyst for progress, a catalyst for change,
Holmium, you embody the spirit of the brave.
    Holmium, the guardian of the unseen,
In your essence, knowledge convenes.
With every breakthrough, a new era unfurls,
In the grand tapestry of Holmium, where wisdom swirls.

# EIGHTEEN

# TREASURE TO ADORE

In the realm of elements, a hidden gem,
Holmium, your presence a rare emblem.
Magnetic allure, a force so strong,
In your atomic dance, a melody is born.
    Oh Holmium, conductor of magnetic might,
You navigate the currents, shining bright.
From lasers to data storage, you excel,
Unraveling the secrets, where wonders dwell.
    In laboratories, scientists explore,
Your properties, a treasure to adore.
Advancing technology, pushing the brink,
Holmium, you're the catalyst, don't shrink.
    In medical realms, a healing hand you lend,
Enhancing imaging, a gift to extend.

Diagnosing ailments with precision's grace,
Holmium, your impact, they embrace.
    Holmium, the guardian of knowledge's door,
In your essence, wisdom does pour.
With every discovery, a new horizon unfolds,
In the grand narrative of Holmium, untold.

# NINETEEN

# INSPIRE WITH YOUR COMPLEXITY

Holmium, an element of hidden allure,
With secrets that science seeks to ensure.
In the realm of magnets, you reign supreme,
A magnetic force, a captivating dream.

Your atomic structure, a symphony divine,
Guiding researchers, their knowledge to refine.
In the pursuit of understanding your might,
Holmium, you shine like a celestial light.

From lasers to technology's advance,
Your presence leaves an indelible stance.
In fiber optics, you weave the way,
Transporting information without delay.

Holmium, a guardian of energy's flow,
Unleashing potentials, letting them grow.

In the vast universe of elements, you stand,
A catalyst, captivating minds and hands.
    Oh Holmium, element of mystery profound,
In your essence, the universe is crowned.
A token of nature's ingenuity,
Holmium, you inspire with your complexity.

# TWENTY

# SCIENTISTS INVESTIGATE

Holmium, a rare gem, hidden in the core,
In the realm of elements, you do adore.
With magnetic allure, you captivate,
Unveiling secrets, as scientists investigate.

    Your atomic dance, a symphony of might,
Guiding research, shedding radiant light.
In lasers and crystals, your power shines,
Pushing boundaries, exploring new confines.

    Holmium, conductor of electromagnetic song,
In your presence, innovation throngs.
From medical imaging to energy's quest,
You fuel progress, surpassing the rest.

    A guardian of knowledge, you stand tall,
Unraveling mysteries, both big and small.

In your magnetic grip, possibilities ignite,
Holmium, forever championing the fight.
  Oh Holmium, element of infinite grace,
In your essence, science finds its place.
A testament to the wonders of the unseen,
Holmium, you're the catalyst of dreams.

# TWENTY-ONE

# GUARDIAN OF DREAMS

Holmium, a gem in the periodic array,
A symphony of electrons in elegant display.
In the realm of magnetism, you reign supreme,
Unleashing forces like a mystical stream.

Your magnetic prowess, a captivating dance,
In lasers and sensors, you cast your trance.
Guiding technology with precision and grace,
Holmium, you leave an indelible trace.

From medical imaging to scientific delight,
You illuminate the path, shining so bright.
A guardian of knowledge, a beacon of light,
Holmium, you lead us through the darkest night.

In the tapestry of elements, you're unique,
A symbol of strength, both bold and sleek.

Your atomic structure, a marvel to behold,
Unveiling secrets, untangling stories untold.

Holmium, conductor of electromagnetic might,
In your presence, the world shines so bright.
A catalyst for progress, a source of inspiration,
Holmium, you fuel innovation's elation.

Oh Holmium, element of infinite grace,
In your essence, science finds its embrace.
A testament to the wonders of the unseen,
Holmium, you're the guardian of dreams.

# TWENTY-TWO

# FUTURE SO BRIGHT

Holmium, a hidden gem of the periodic table,
Your presence quietly resonates, a fable.
In laboratories, scientists seek your might,
Unveiling mysteries, like stars in the night.

With magnetic allure, you captivate the gaze,
Guiding discoveries, in so many ways.
In lasers and optics, you find your domain,
Harnessing energy, like a celestial flame.

Oh Holmium, conductor of electromagnetic symphony,
You orchestrate marvels, shaping our destiny.
From medical realms to sustainable power,
You empower progress, hour after hour.

Holmium, the guardian of knowledge's gate,
In your essence, secrets illuminate.

With each breakthrough, a new era unfolds,
In the grand tapestry of Holmium, wisdom beholds.
 A testament to the wonders of the unseen,
Holmium, you stand tall, forever serene.
A catalyst for innovation, a symbol of might,
Holmium, you guide us towards a future so bright.

# TWENTY-THREE

# FIBER OPTICS TO LASERS

In the realm of elements, a jewel does reside,
Holmium, a marvel, where wonders coincide.
With magnetic might, you command the stage,
Guiding innovation, fueling progress with rage.

Oh Holmium, conductor of electromagnetic dreams,
In your presence, science sings in vibrant streams.
From fiber optics to lasers that gleam,
You illuminate the path, like a cosmic beam.

In medical marvels, your touch is profound,
Enhancing diagnostics, where hope is found.
MRI machines, your magnetic embrace,
Revealing the mysteries, with each scanning trace.

Holmium, a guardian of knowledge's door,
Unraveling secrets, you forever explore.

In laboratories, your power is unveiled,
Pushing boundaries, where discoveries are hailed.
   Oh Holmium, element of infinite grace,
In your essence, science finds a sacred space.
A catalyst for progress, a beacon so bright,
Holmium, you guide us towards the light.

# TWENTY-FOUR

# HEALING THE SOUND

Holmium, a symphony of atomic might,
Your presence shines with an ethereal light.
In the realm of elements, you stand tall,
Unveiling mysteries, captivating all.

From lasers to optical fibers, you lead,
Harnessing energy, fulfilling every need.
Your magnetic allure, a force so strong,
Guiding innovation, inspiring all along.

Holmium, conductor of electromagnetic dreams,
A catalyst for progress, it seems.
In medical realms, your touch is profound,
Enhancing diagnostics, healing the sound.

In laboratories, scientists explore,
Your properties, a treasure to adore.
Advancing technology, pushing the brink,
Holmium, you're the catalyst, don't shrink.

Oh Holmium, element of infinite grace,
In your essence, science finds its place.
A testament to the wonders of the unseen,
Holmium, you inspire, forever serene.

# TWENTY-FIVE

# MYSTERY PROFOUND

Holmium, oh element of rare delight,
In the realm of science, you shine so bright.
With atomic power and magnetic charm,
You captivate minds, setting off the alarm.

In optical fibers, you pave the way,
Carrying information without delay.
A conductor of light, you guide it through,
Enabling connections, both old and new.

Holmium, the guardian of energy's flow,
Unleashing potentials, letting them grow.
In the world of lasers, you take the lead,
Harnessing photons, fulfilling every need.

Oh Holmium, element of mystery profound,
In your essence, the universe is crowned.
A symbol of innovation, you inspire,
Pushing boundaries higher and higher.

Exploring your secrets, scientists strive,
To unravel the wonders you contrive.
Holmium, a catalyst for progress and more,
You're an element we truly adore.

# TWENTY-SIX

# DIVINE

In the realm of elements, a jewel does shine,
Holmium, a marvel, a treasure so fine.
With atomic number sixty-seven you stand,
A beacon of science, in the periodic land.

In the magnetic field, you wield your might,
Aligning spins, creating a wondrous sight.
Your magnetic properties, a phenomenon rare,
Exploring your nature, scientists dare.

Holmium, conductor of electromagnetic dreams,
In lasers and optics, your brilliance gleams.
From infrared to visible, you emit a vibrant hue,
Guiding innovation, unveiling what's true.

Your presence in medicine is nothing short of profound,
Enhancing imaging, healing, and sound.

MRI machines, with your magnetic embrace,
Revealing the inner workings of the human race.

Oh Holmium, element of infinite grace,
In your essence, science finds a sacred space.
A catalyst for progress, a symbol of might,
Holmium, you guide us towards the light.

Through your unique properties, we learn and grow,
Unraveling the mysteries only you can show.
Holmium, a testament to nature's grand design,
Forever captivating, forever divine.

# TWENTY-SEVEN

# SCIENCE REDEEMS

In the realm of elements, you shine bright,
Holmium, a marvel, a captivating sight.
With magnetic allure, you command the stage,
Guiding innovation, turning the page.

    In lasers, your power is unleashed,
A symphony of light, a masterpiece.
From red to green, your colors dance,
Illuminating paths, taking a chance.

    Holmium, conductor of electromagnetic dreams,
Through your essence, science redeems.
In the world of medicine, you play a crucial role,
Enhancing diagnostics, healing the soul.

    In laboratories, you're a treasure to explore,
Unraveling secrets, opening new doors.

A catalyst for progress, a beacon so clear,
Holmium, you inspire, casting away fear.
   Oh Holmium, element of infinite grace,
In your presence, science finds its place.
A testament to nature's symphony divine,
Holmium, you shine, forever entwined.

# TWENTY-EIGHT

# DISCOVERIES GALORE

Holmium, the element of strength and might,
In the realm of science, you shine so bright.
With your magnetic power, you hold the key,
To unlock the secrets of the universe, you see.

In lasers and optics, your brilliance beams,
Guiding innovation, fulfilling our dreams.
A conductor of light, you lead the way,
Navigating the realms of technology each day.

In medical fields, you make a profound mark,
Enhancing diagnostics, leaving a lasting spark.
MRI machines, with your magnetic allure,
Revealing the hidden depths, so pure.

Oh Holmium, element of infinite grace,
In your presence, knowledge finds its place.
A catalyst for progress, a symbol of hope,
Inspiring scientists, helping them cope.

In laboratories, your mysteries unfold,
Unveiling wonders, yet to be told.
Holmium, you're a treasure to explore,
Opening doors to discoveries galore.

So let us celebrate this element divine,
Holmium, forever radiant, forever will shine.

# TWENTY-NINE

# SPECTRAL DANCE

Holmium, a jewel in the periodic table's crown,
With properties unique, you never let us down.
In the world of magnets, you reign supreme,
Guiding energy, like a mesmerizing dream.

Your magnetic field, a force to behold,
Pulling us closer, in a tale untold.
From headphones to electric motors, you power,
Enchanting us with your electromagnetic tower.

Oh Holmium, element of captivating allure,
In your presence, innovation finds its cure.
A conductor of energy, a catalyst for change,
You weave through circuits, a symphony so strange.

In lasers, you dazzle with a spectral dance,
Painting the universe with a vibrant expanse.

From red to blue, your colors intertwine,
Shaping the future, with a light so divine.

Holmium, a testament to science's quest,
Unveiling mysteries, putting theories to the test.
Oh, how we marvel at your atomic might,
Holmium, forever shining, forever in sight.

# THIRTY

## ATOMIC ART

In the realm of chemistry, a gem we find,
A rare element of a distinctive kind.
Holmium, you stand with grace and might,
A beacon of wonder, shining so bright.
    Within magnets, your power lies,
Pulling forces, secrets in disguise.
From generators to electric cars,
You energize the world, reaching for the stars.
    Your atomic structure, so elegantly arranged,
Captivating scientists, deeply engaged.
In the lab, your secrets unfold,
Unveiling mysteries, untold and bold.
    Holmium, conductor of vibrant light,
You illuminate the world, day and night.

From lasers to fiber optics, your domain,
Guiding communications, breaking the chain.

Oh Holmium, element of boundless might,
You fill our hearts with awe and delight.
A symbol of progress, a catalyst profound,
Holmium, you inspire, the world around.

So let us celebrate your atomic art,
Holmium, forever etched in science's heart.
In laboratories, your legacy will thrive,
As we continue to explore and strive.

# THIRTY-ONE

# CORNERS OF THE EARTH

In the realm of elements, you stand tall,
Holmium, a wonder, captivating all.
With atomic number sixty-seven you reside,
A symbol of knowledge, impossible to hide.

Oh Holmium, your properties so rare,
Magnetic and luminescent, beyond compare.
In lasers, you emit a vibrant light,
Guiding us through darkness, shining bright.

From green to red, your colors unfold,
A mesmerizing spectacle, a story to be told.
In laboratories, scientists seek your grace,
Unlocking secrets of the atomic space.

Holmium, conductor of energy and power,
In medicine, you prove to be a tower.

MRI machines, with your magnetic might,
Reveal the inner workings, shedding light.
    Oh Holmium, element of infinite worth,
You bring progress to the corners of the Earth.
A symbol of innovation, a beacon of hope,
Holmium, in our hearts, you forever elope.

# THIRTY-TWO

# METAL WITH ALLURE

Holmium, a metal with allure so rare,
Your presence in the periodic table, beyond compare.
Magnetic and luminescent, you captivate,
Unveiling secrets that scientists appreciate.

In the realm of lasers, you take the lead,
Your vibrant light, a source of awe indeed.
From green to red, you paint the spectrum wide,
Guiding humanity with a radiant stride.

In the world of medicine, you play a crucial role,
Enhancing diagnostics, healing the soul.
MRI machines, with your magnetic might,
Reveal the hidden depths, shining a guiding light.

Oh Holmium, element of wonder and grace,
Your contributions to science, we embrace.
A catalyst for progress, a symbol of hope,
Inspiring generations, helping us cope.

In laboratories, your mysteries unfold,
Unraveling the secrets that were once untold.
Holmium, you're a treasure to explore,
Opening doors to knowledge, forevermore.
So let us celebrate this element divine,
Holmium, forever radiant, forever shall shine.

# THIRTY-THREE

# LUMINESCENT HUE

Holmium, a gem in the periodic chart,
A rare beauty, a work of scientific art.
With atomic number sixty-seven you stand,
A symbol of power, a gift from nature's hand.

In magnets, you hold a magnetic force,
Attracting and repelling, staying on course.
From motors to headphones, you lend your might,
Electrifying our world, shining so bright.

Oh Holmium, element of magnetic grace,
You permeate our lives, leaving a trace.
In lasers, you emit a vibrant glow,
Harnessing light, putting on a show.

In laboratories, you captivate the minds,
Unraveling mysteries, breaking through binds.

Holmium, you inspire, you ignite the flame,
Fueling innovation, forever changing the game.
    So let us celebrate your atomic song,
Holmium, forever in science, you belong.
Through your magnetic charm and luminescent hue,
You guide us towards progress, forever anew.

# THIRTY-FOUR

# SYMBOL OF INNOVATION

Oh Holmium, element of rare and precious worth,
In the realm of science, you hold immense berth.
Your atomic number fifty-nine, a symbol of might,
You illuminate the path, casting away the night.

Magnetic in nature, you dance with grace,
Attracting and repelling, leaving a trace.
From MRI machines to laser devices,
Your presence empowers, your essence entices.

In laboratories, your secrets unfurl,
Unveiling wonders, like a precious pearl.
Holmium, conductor of energy and light,
You guide our discoveries, shining so bright.

Oh, element of Holmium, you're a marvel to behold,
Your luminescent glow, a story yet untold.

In the world of technology, you lead the way,
Igniting progress, with each passing day.

So let us celebrate your atomic art,
Holmium, forever etched in science's heart.
From the depths of the periodic table, you rise,
A symbol of innovation, forever in our eyes.

# ABOUT THE AUTHOR

Walter the Educator is one of the pseudonyms for Walter Anderson. Formally educated in Chemistry, Business, and Education, he is an educator, an author, a diverse entrepreneur, and he is the son of a disabled war veteran. "Walter the Educator" shares his time between educating and creating. He holds interests and owns several creative projects that entertain, enlighten, enhance, and educate, hoping to inspire and motivate you.

Follow, find new works, and stay up to date
with Walter the Educator™
at WaltertheEducator.com

www.ingramcontent.com/pod-product-compliance
Lightning Source LLC
LaVergne TN
LVHW052001060526
838201LV00059B/3763